Problems and Solutions Manual

FATIGUE DESIGN

Life Expectancy of Machine Parts

Eliahu Zahavi
with Vladimir Torbilo

PROBLEM 1 (Chapter 3)

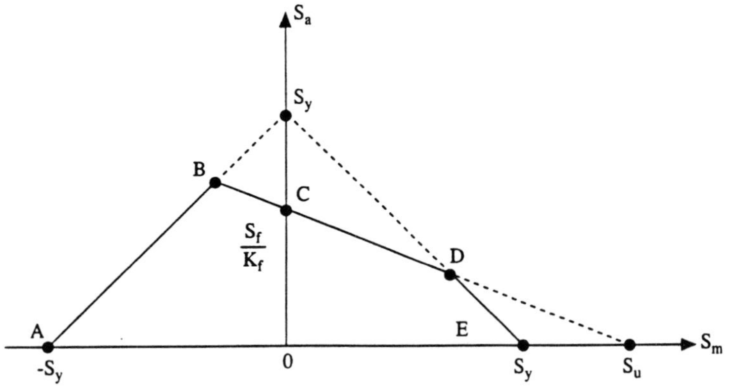

Figure P.1

Figure P.1 above, shows a limit diagram of a machine part in S_a vs. S_m coordinates.
Required: Draw an equivalent diagram in S_{max}, S_{min} vs. S_m coordinates.

SOLUTION

The approach is to locate the equivalent points A, B, C, D, and E. The diagram is drawn connecting the respective points with straight lines.

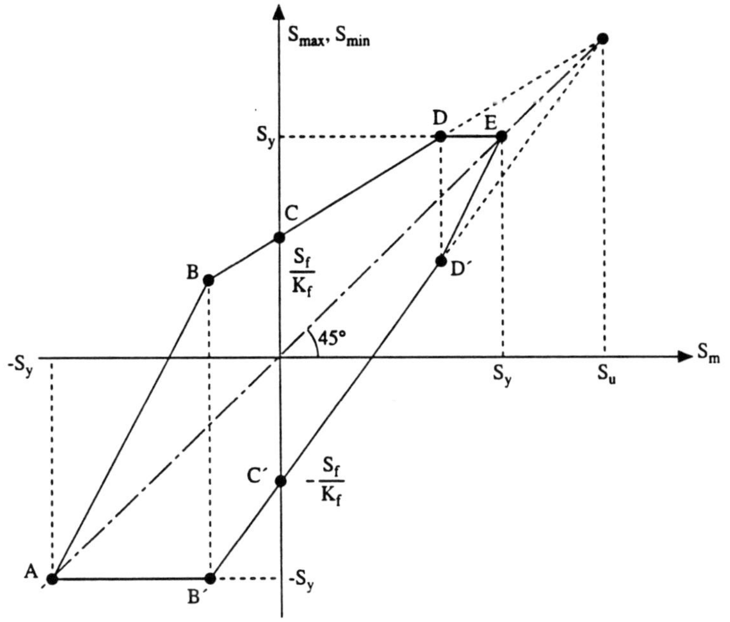

Figure P.2

PROBLEM 2 (Chapter 3)

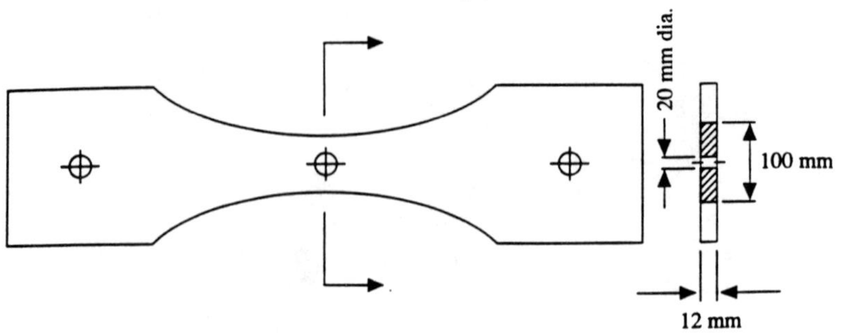

Figure P.3

The link shown in Figure P.3 above is made of SAE 1020 steel, hot rolled and machined according to specifications. The part is prestressed at the assembly by an axial tensional load of 80,000 N and subject to a reversing axial force of ± 42,000 N, superimposed on the prestressing load.

Compute the factor of safety.

SOLUTION

The nominal stresses in the link due to the loading are as follows:

mean stress

$$S_m = \frac{F_m}{A} = \frac{80000}{(0.100 - 0.02) \times 0.012} \times 10^{-6} = 83.33\,\text{MPa}$$

alternating stress amplitude

$$S_a = \frac{F_a}{A} = \frac{42,000}{(0.100 - 0.02) \times 0.012} \times 10^{-6} = 43.75\,\text{MPa}$$

The material properties are as follows:

ultimate tensile strength

$$S_u = 440\,\text{MPa}\ (64\,\text{ksi})$$

yield point

$$S_y = 262\,\text{MPa}$$

(Note: in accordance with the stress method, a monotonic yield point is used.)

The fatigue limit is computed from Equation (3.14). The fatigue limit modification factors, based on Equations (3.9) and (3.10) and Figure 3.15 equal:

$$K_a = 0.9, \qquad K_b = 1.0, \qquad K_c = 0.85$$

The fatigue limit becomes:

$$S_f = 0.9 \times 1.0 \times 0.85 \times \frac{440}{2} = 168.3\,\text{MPa}$$

The fatigue notch factor K_f is computed from Equation (3.22). The theoretical stress concentration factor from Figure 3.30 is:

$$K_t = 2.52$$

The notch sensitivity q is computed from Equation (3.21) and Figure 3.18. The Neuber constant is:

$$a = 0.25\,\text{mm}$$

and the notch sensitivity is:

$$q = \frac{1}{1 + \sqrt{\dfrac{0.25}{10}}} = 0.863$$

The fatigue notch factor becomes:

$$K_f = 1 + 0.863(2.52 - 1) = 2.31$$

To compute the factor of safety, one can choose:

(a) Equations (3.43) and (3.44) or
(b) Equations (3.49) and (3.50).

(a) The following expression is taken based on Figure 3.44

$$\frac{S_f / K_f}{S_{\alpha,\lim}} = \frac{S_u}{S_u - S_m}$$

From this, one obtains:

$$S_{\alpha,\lim} = \frac{S_f}{K_f}\frac{S_u - S_m}{S_u}$$

$$= \frac{168.3}{2.31}\frac{440 - 83.33}{440} = 59.09\,\text{MPa}$$

The safety factor *FS* is based on Equation (3.43) and *FS'* is based on Equation (3.44):

$$FS = \frac{59.09}{43.75} = 1.35$$

$$FS' = \frac{262}{83.33 + 43.75} = 2.06$$

Choosing the lower value, the correct safety factor is:

$$FS = 1.35$$

(b) When the prestressing load F_m is of uncertain value, one should consider Equation (3.49) in lieu of (3.43):

$$\frac{1}{FS} = \frac{83.33}{440} + 2.31\frac{43.75}{168.3}, \qquad FS = 1.27$$

A complementary safety factor per Equation (3.50):

$$FS' = 2.06$$

The safety factor using Equations (3.49) and (3.50) becomes:

$$FS = 1.27$$

Conclusion: Given an accurate prestressing load, F_m, the safety factor is 1.35, compared with a conservative estimate of 1.27.

PROBLEM 3 (Chapter 3)

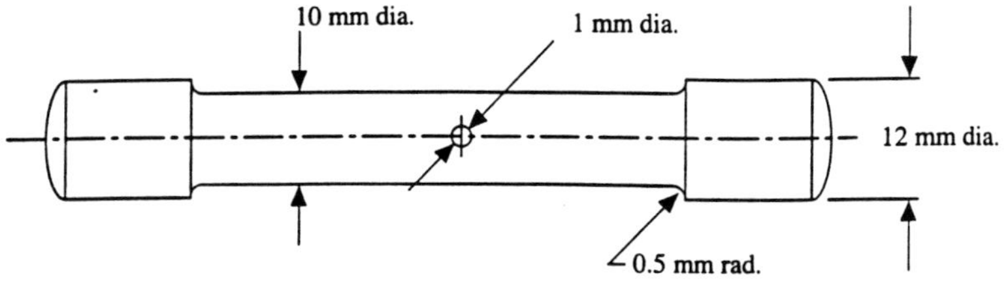

Figure P.4

The part shown in Figure P.4 above is made of steel SAE 1045, BHN = 225, by machining. Compute and draw an *S–N* diagram for the shown part in reversed loading.

SOLUTION

The ultimate strength of the material is (see Table A.1):

$$S_u = 725\,\text{MPa} \ \ (105\,\text{ksi})$$

The fatigue limit modification factors are (See Equations (3.9) and (3.10) and Figure 3.15):

$$K_a = 0.9, \qquad K_b = 1.0, \qquad K_c = 0.81$$

The fatigue limit equals:

$$S_f = 0.9 \times 1.0 \times 0.81 \times \frac{725}{2} = 264.3\,\text{MPa}$$

For stress concentration, three locations require an investigation: at the right fillet, left fillet, and next to the center hole.

Consider the fillets first. Based on:

$$D/d = 1.2, \qquad r/d = 0.05$$

and Figure 3.27:

$$K_t = 2.05$$

Consider the center hole. Here:

$$d/D = 0.1$$

and

$$K_t = 2.75$$

The latter factor is higher, therefore it is used for further analysis. From Figure 3.19 the notch sensitivity equals:

$$q = 0.7$$

The fatigue notch factor becomes:

$$K_f = 1 + 0.7(2.75 - 1) = 2.23$$

The fatigue strength at 10^6 cycles equals

$$\frac{S_f}{K_f} = \frac{264.3}{2.23} = 118.5\,\text{MPa}$$

The fatigue strength at 10^3 cycles equals

$$0.9\,S_u = 0.9 \times 725 = 652.5\,\text{MPa}$$

The complete *S–N* curve is shown in Figure P.5.

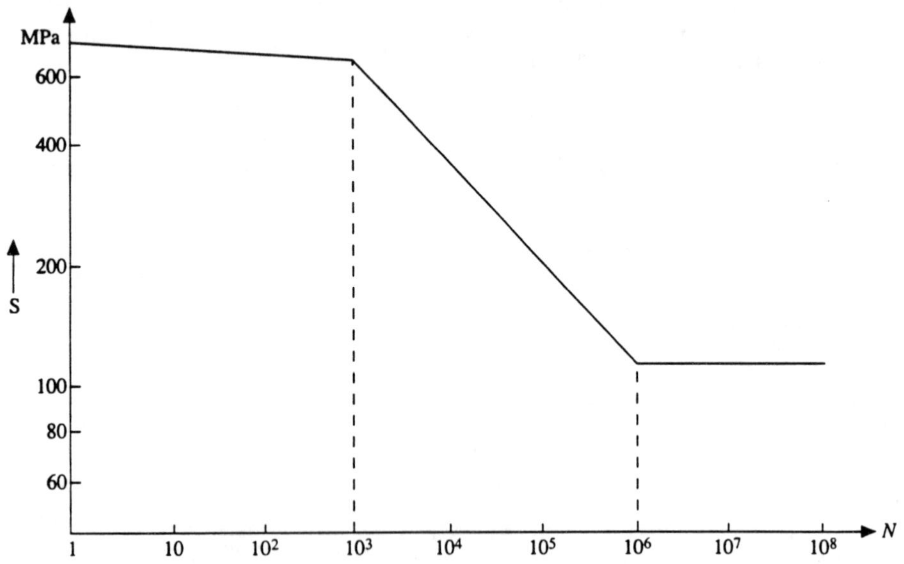

Figure P.5

PROBLEM 4 (Chapter 3)

(Use the data given in Problem 3)
For the part described in Problem 3, compute and draw:

- **(a)** a complete limit diagram;
- **(b)** a constant life diagram.

SOLUTION

From the Solution of Problem 3:

$$S_u = 725\,\text{MPa}$$

$$\frac{S_f}{K_f} = 118.5\,\text{MPa}$$

(a) From Table A.1 the yield point for steel SAE 1045, BHN = 225 is:

$$S_y = 634\,\text{MPa}$$

We draw the requested limit diagram, as shown in Figure P.6. (Note: in accordance with the stress method, the monotonic yield point is used.)

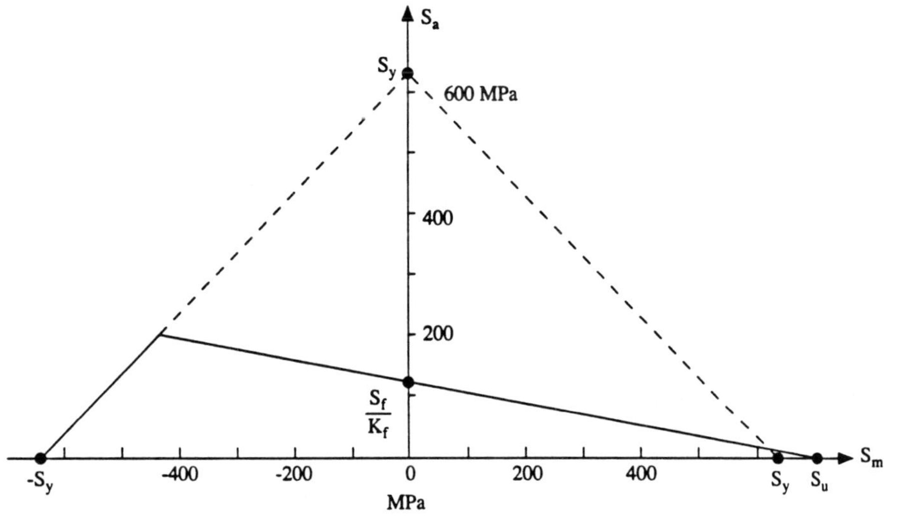

Figure P.6

(b) The constant life diagram is produced from the same data utilizing fatigue strength values from Figure P.5:

at $N = 10^3$ $\quad S = 652.5\,\text{MPa}$

at $N = 10^4$ $\quad S = 340\,\text{MPa}$

at $N = 10^5$ $\quad S = 200\,\text{MPa}$

The diagram is shown in Figure P.7.

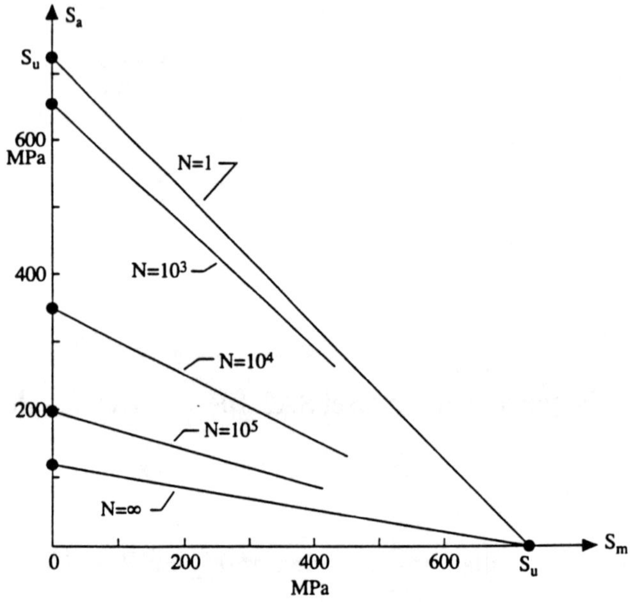

Figure P.7

PROBLEM 5 (Chapter 3)

The link presented in Problem 2 is prestressed by a load of 150,000 N, subject to a reversing force of ± 60,000 N. Compute the life expectancy of the link by the stress method.

SOLUTION

The nominal stesses are as follows:

mean stress:

$$S_m = \frac{150{,}000}{(0.100 - 0.02) \times 0.012} \times 10^{-6} = 156.25\,\text{MPa}$$

alternating stress amplitude:

$$S_a = \frac{60{,}000}{(0.100 - 0.02) \times 0.012} \times 10^{-6} = 62.5\,\text{MPa}$$

The ultimate tensile strength is:

$$S_u = 440\,\text{MPa}$$

The equivilent alternating stress, based on Equation (3.63) equals

$$S_N = 440 \times \frac{62.5}{440 - 156.25} = 96.9\,\text{MPa}$$

The fatigue limit and the fatigue notch factor taken from the solution of Problem 2 are:

$$S_f = 168.3\,\text{MPa}$$

$$K_f = 2.31$$

The life expectancy, *N*, based on Equation (3.60) is:

$$N = 1000\left(\frac{0.9 \times 440}{96.9}\right)^{\frac{3}{\log\frac{0.9\times440\times2.31}{168.3}}} = 312{,}346 \text{ cycles}$$

PROBLEM 6 (Chapter 3)

A shaft of 50 mm diameter is made of SAE 1144 steel with a hardness of 265 BHN. The shaft is machined. The stress concentration factor, at its critical section, is $K_t = 1.7$. The notch sensitivity is $q = 0.86$. The shaft, at its critical section, is subject to nominal bending stresses, varying as follows:

S = 135 to 480 MPa during 89% of revolutions,
S = 100 to 515 MPa during 10% of revolutions,
S = 70 to 550 MPa during 1% of revolutions.

Compute the life expectancy of the shaft.

SOLUTION

The ultimate tensile strength of the material, per Table A.1, is:

$$S_u = 930\,\text{MPa}\ (135\,\text{ksi})$$

The fatigue limit modification factors are:

$$k_a = 1.0, \qquad k_b = 0.85, \qquad k_c = 0.8$$

The fatigue limit is:

$$S_f = 1.0 \times 0.85 \times 0.8 \times \frac{930}{2} = 316.2\,\text{MPa}$$

The fatigue notch factor equals:

$$K_f = 1 + 0.86(1.7 - 1) = 1.60$$

In order to obtain the final life expectancy, the computation should follow three stress levels. At the first stress level, the nominal mean stress and alternating stress amplitude are:

$$S_m = 0.5(480 + 135) = 307.5\,\text{MPa}$$
$$S_a = 0.5(480 - 135) = 172.5\,\text{MPa}$$

The resulting equivalent alternating stress equals:

$$S_N = 930\frac{172.5}{930 - 307.5} = 257.7\,\text{MPa}$$

while the life expectancy at this, first, stress level is:

$$N_1 = 1000\left(\frac{0.9 \times 930}{257.7}\right)^{\frac{3}{\log\frac{0.9\times930}{316.2/1.6}}} = 281{,}000\,\text{cycles}$$

At the second stress level, the mean and the alternating stresses are:

$$S_m = 0.5(515 + 100) = 307.5\,\text{MPa}$$
$$S_a = 0.5(515 - 100) = 207.5\,\text{MPa}$$

and the resulting equivalent alternating stress is:

$$S_N = 930\frac{207.5}{930-307.5} = 310\,\text{MPa}$$

The life expectancy at the second stress level becomes:

$$N_2 = 1000\left(\frac{0.9\times 930}{310}\right)^{\frac{3}{\log\frac{0.9\times 930}{316.2/1.6}}} = 116{,}000\,\text{cycles}$$

At the third stress level the stresses are:

$$S_m = 0.5(550+70) = 310\,\text{MPa}$$
$$S_a = 0.5(550-70) = 240\,\text{MPa}$$

and the resulting equivalent alternating stress equals:

$$S_N = 930\frac{240}{930-310} = 360\,\text{MPa}$$

The life expectancy corresponding to the third level becomes:

$$N_3 = 1000\left(\frac{0.9\times 930}{360}\right)^{\frac{3}{\log\frac{0.9\times 930}{316.2/1.6}}} = 56{,}700\,\text{cycles}$$

Using the computed life expectancies at the respective stress levels and by means of the Palmgren-Miner rule, Equation (3.89), the expression becomes:

$$\frac{0.89}{281{,}000} + \frac{0.10}{116{,}000} + \frac{0.01}{56{,}700} = \frac{1}{N}$$

From this results the final life expectancy:

$$N = 237{,}800\,\text{cycles}$$

PROBLEM 7 (Chapter 4)

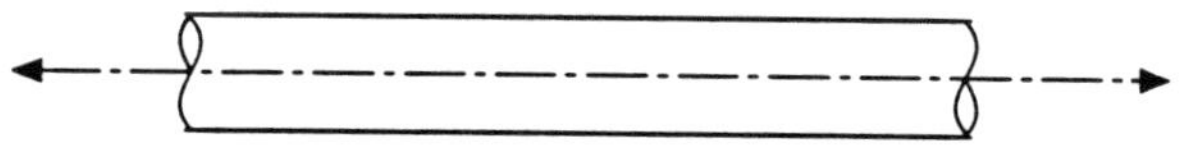

Figure P.8

A smooth uniaxial part, see Figure P.8 above, is subject to a reversing axial load. The part is made of SAE 1020 steel, hot rolled.

Compute and draw a diagram of strain versus life. What will be the number of load reversals to failure if the strain amplitude will equal 0.01?

SOLUTION

The following data is taken from Table A.1:

fatigue strength coefficient	$\sigma'_f = 895\,\text{MPa}$
fatigue ductility coefficient	$\varepsilon'_f = 0.41$
fatigue strength exponent	$b = -0.12$
fatigue ductility exponent	$c = -0.51$
modulus of elasticity	$E = 203{,}000\,\text{MPa}$

The diagram of strain versus life is drawn by superimposing the line of plastic strain on the line of elastic strain, see Figure 4.25.
The following expression for elastic strain is based on Equation (4.28):

$$\frac{\Delta\varepsilon_e}{2} = \frac{895}{203{,}000}\left(2N_f\right)^{-0.12} = 0.0044\left(2N_f\right)^{-0.12}$$

while the expression of plastic strain is based on Equation (4.29):

$$\frac{\Delta\varepsilon_p}{2} = 0.41\left(2N_f\right)^{-0.51}$$

The resulting correlation for the strain versus life is:

$$\frac{\Delta\varepsilon}{2} = 0.0044\left(2N_f\right)^{-0.12} + 0.41\left(2N_f\right)^{-0.51}$$

The corresponding diagram is shown in Figure P.9.

With strain amplitude of 0.01, the life is:

$2N_f = 2000$ reversals

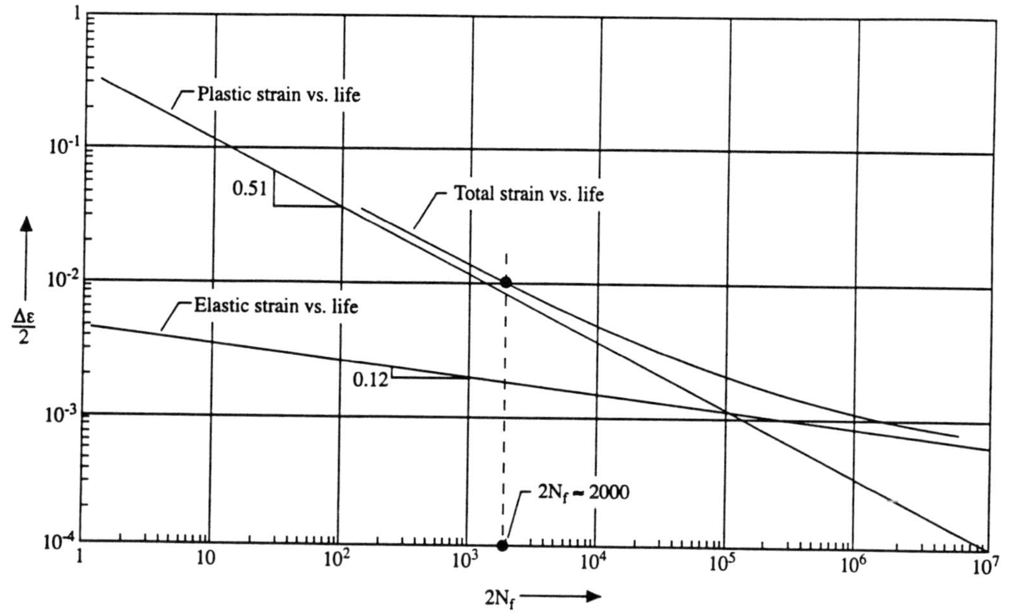

Figure P.9

PROBLEM 8 (Chapter 4)

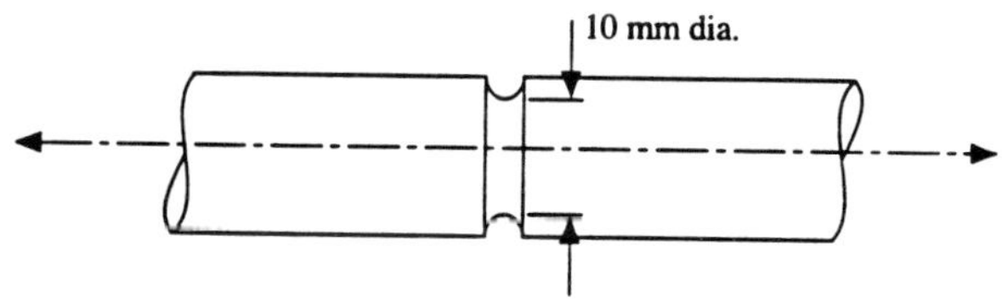

Figure P.10

The part shown in Figure P.10 above is made of SAE 1020 steel, hot rolled and machined with a circumferential groove that causes a stress concentration factor $K_t = 2.8$. The part is subject to a reversing axial load of ± 20,000 N. Compute the life expectancy by the strain method.

SOLUTION

The following properties are taken from Table A.1:

modulus of elasticity	$E = 203{,}000\,\text{MPa}$
fatigue strength coefficient	$\sigma'_f = 895\,\text{MPa}$
fatigue ductility coefficient	$\varepsilon'_f = 0.41\,\text{MPa}$
cyclic strain hardening exponent	$n' = 0.18$

The true stress-strain correlation as per Equation (4.21) is:

$$\frac{\Delta\varepsilon}{2} = \frac{1}{203{,}000} \times \frac{\Delta\sigma}{2} + 0.41\left(\frac{1}{895} \times \frac{\Delta\sigma}{2}\right)^{\frac{1}{0.18}}$$

This expression is plotted in Figure P.11 and shown as line OA. The nominal alternating engineering stress amplitude in the critical section is:

$$\frac{\Delta S}{2} = \frac{20{,}000}{\frac{\pi}{4} \times 0.01^2} \times 10^{-6} = 254.6 \text{ MPa}$$

The nominal alternating engineering strain amplitude is:

$$\frac{\Delta e}{2} = \frac{254.6}{203{,}000} = 0.00125$$

The Neuber rule, Equation (4.39) is expressed as:

$$\frac{\Delta\varepsilon}{2}\frac{\Delta\sigma}{2} = 2.8^2 \times 254.6 \times 0.00125 = 2.495$$

The above expression is plotted in Figure P.11 and shown as line XY. At the crossing point of lines OA and XY (point Z) the true strain amplitude is:

$$\frac{\Delta\varepsilon}{2} = 0.0066$$

As per Figure P.9, the number of load reversals to failure is:

$$2N_f = 6000$$

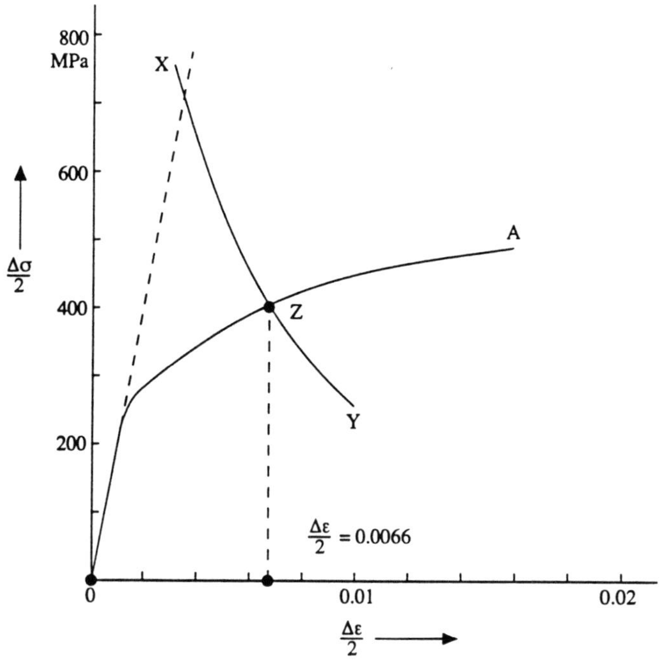

Figure P.11

PROBLEM 9 (Chapter 4)

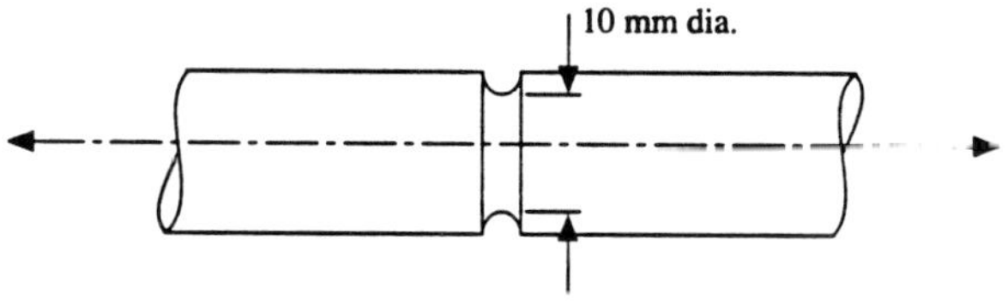

Figure P.12

The part shown in Figure P.12 above has circumferential groove with a concentration factor K_t = 2.98. The part is subject to a reversing axial load of ± 38,000 N. Two materials are considered:

(a) SAE 1020, BHN = 108
(b) SAE 4142, BHN = 380.

Which of the two metals provides the higher life expectancy?

SOLUTION

The solution is based on the strain theory.

(a) SAE 1020. The properties are as follows:

modulus of elasticity	$E = 203{,}000\,\text{MPa}$
fatigue strength coefficient	$\sigma'_f = 895$ MPa
fatigue ductility coefficient	$\varepsilon'_f = 0.41$
fatigue strength exponent	$b = -0.12$
fatigue ductility exponent	$c = -0.51$
cyclic strain hardening exponent	$n' = 0.18$

The resulting true stress-strain correlation, Equation (4.21), becomes:

$$\frac{\Delta\varepsilon}{2} = \frac{895}{203{,}000}\frac{\Delta\sigma}{2} + 0.41\left(\frac{1}{895} \times \frac{\Delta\sigma}{2}\right)^{\frac{1}{0.18}}$$

which is plotted in Figure P.13. The strain versus life is expressed by Equation (4.31) which takes the form:

$$\frac{\Delta\varepsilon}{2} = \frac{895}{203{,}000}\left(2N_f\right)^{-0.12} + 0.41\left(2N_f\right)^{-0.51}$$

The expression is plotted in Figure P.14. The nominal alternating engineering stress amplitude in the critical section equals:

$$\frac{\Delta S}{2} = \frac{38{,}000}{\frac{\pi}{4} \times 0.01^2} \times 10^{-6} = 483.83 \text{ MPa}$$

and the nominal alternating engineering strain amplitude:

$$\frac{\Delta e}{2} = \frac{483.83}{203{,}000} = 0.0023834$$

Using the Neuber rule, Equation (4.39) takes the form:

$$\frac{\Delta\varepsilon}{2}\frac{\Delta\sigma}{2} = 2.98^2 \times 483.83 \times 0.0023834 = 10.26$$

The above expression plotted in Figure P.13 produces a hyperbole which crosses the SAE 1020 stress-strain curve at point A. Consequently, the true strain equals:

$$\frac{\Delta\varepsilon}{2} = 0.021$$

Using Figure P.14, the following life expectancy is obtained:

$$2N_f = 500 \text{ reversals}$$

(b) SAE 4142. The properties are as follows:

modulus of elasticity	$E = 207{,}000$ MPa
fatigue strength coefficient	$\sigma'_f = 1825$ MPa
fatigue ductility coefficient	$\varepsilon'_f = 0.45$
fatigue strength exponent	$b = -0.08$
fatigue ductility coefficient	$c = -0.75$
cyclic strain hardening exponent	$n' = 0.17$

The true stress-strain correlation has the form:

$$\frac{\Delta\varepsilon}{2} = \frac{1}{207{,}000} \times \frac{\Delta\sigma}{2} + 0.45\left(\frac{1}{1825} \times \frac{\Delta\sigma}{2}\right)^{\frac{1}{0.17}}$$

which is plotted in Figure P.13. The strain versus life is expressed by:

$$\frac{\Delta\varepsilon}{2} = \frac{1825}{207{,}000}\left(2N_f\right)^{-0.08} + 0.45\left(2N_f\right)^{-0.75}$$

which is plotted in Figure P.14. The SAE 4142 stress-strain curve crosses the hyperbole at point B. Consequently the true strain equals:

$$\frac{\Delta\varepsilon}{2} = 0.0115$$

and the life expectancy, as per Figure P.14, is:

$$2N_f = 260 \text{ reversals}$$

Thus, steel SAE 1020 will provide a longer life.

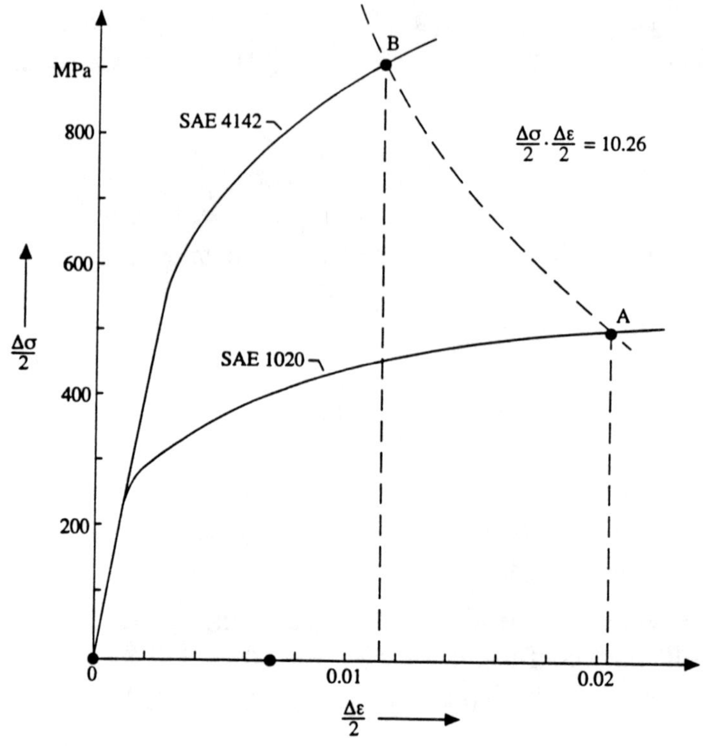

Figure P.13

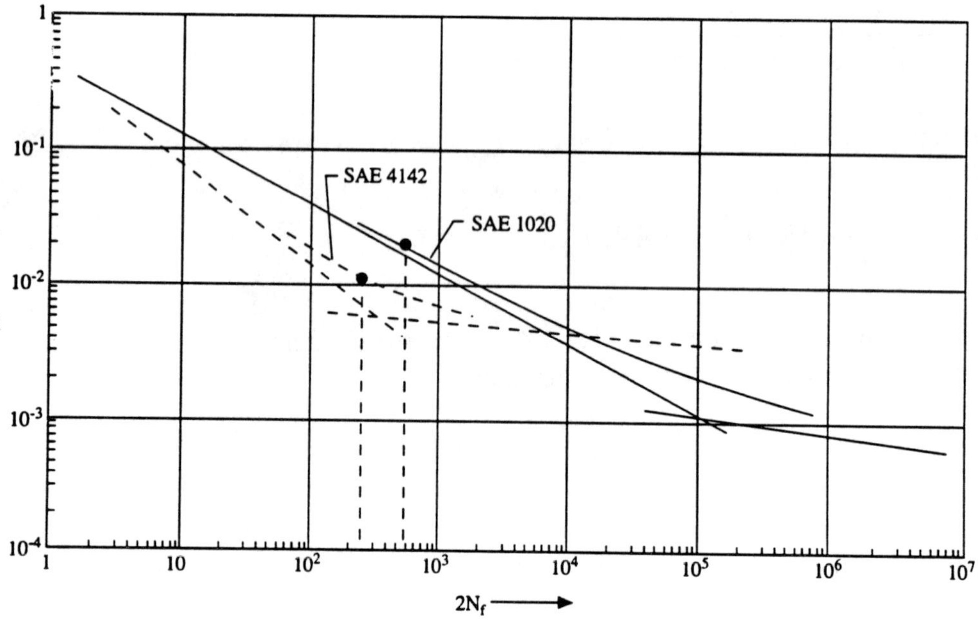

Figure P.14

PROBLEM 10 (Chapter 5)

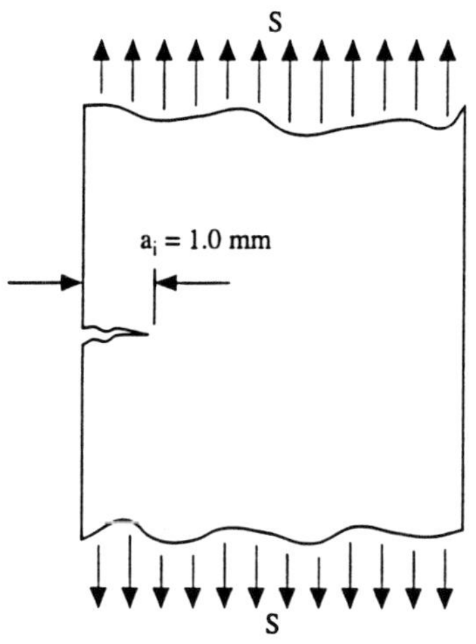

Figure P.15

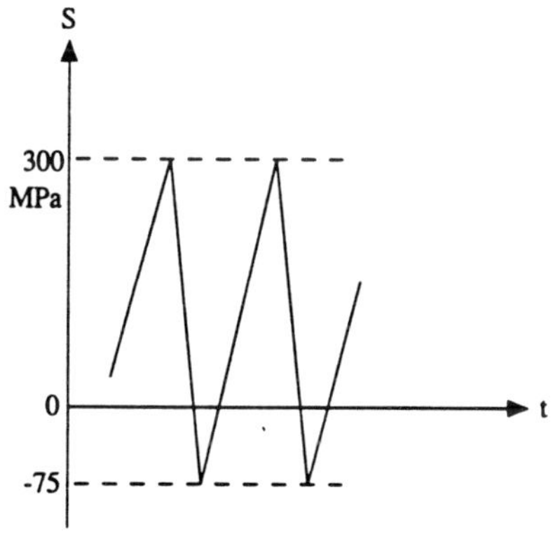

Figure P.16

Figures P.15 and P.16 above represent a flat plate subject to a variable load. It is made of A-441 steel with the following properties:

fracture toughness	$K_{Ic} = 100\,\text{MPa}\sqrt{\text{m}}$
threshold stress intensity factor	$K_{th} = 6.0\ \text{MPa}\sqrt{\text{m}}$
constant factor in Equation (5.85)	$C = 6.9 \times 10^{-12}$
crack growth exponent in Equation (5.85)	$n = 3.0$

The variable load is:

$$S = -75 + 300\ \text{MPa}$$

A small crack of the length

$$a_i = 1.0\,\text{mm} = 0.001\,\text{m}$$

is detected.

Required:

(a) Based on the above predict if there is a danger of crack propagation.
(b) What is the number of cycles to the final fracture?

SOLUTION

To arrive at the proper answer the compression during load fluctuation is ignored and it is assumed that the stress fluctuation has the form:

$$S = 0\,\text{to}\,300\ \text{MPa}$$

(a) At the onset of crack growth, see Figure 5.19, the threshold value of stress intensity factor must equal:

$$K_{th} = 1.12 \cdot \sigma \cdot \sqrt{\pi a} = 1.12 \times 300 \times \sqrt{\pi a}$$

see Equation (5.22). Consequently the initial length of a crack should be:

$$a = \frac{1}{\pi}\left(\frac{6.0}{1.12 \times 300}\right)^2 = 0.0001\ \text{m}$$

The given length is larger, therefore, under the fluctuating loading, the crack will continue to grow.

(b) The initial crack a_i is as shown in Figure P.15. The final crack length is obtained from the following expression:

$$K_{Ic} = 1.12 \cdot \sigma \cdot \sqrt{\pi a_f} = 1.12 \times 300\sqrt{\pi a_f}$$

It follows that:

$$a_f = \frac{1}{\pi}\left(\frac{100}{1.12 \times 300}\right)^2 = 0.02819\ \text{m}$$

The life expectancy of the plate is computed by means of Equation (5.87):

$$N_f = \frac{2}{(3-2)\times 6.9\times 10^{-12}\times \left(1.12\times 300\times \sqrt{\pi}\right)^3}\left(\frac{1}{\sqrt{0.001}}-\frac{1}{\sqrt{0.02819}}\right) = 35{,}221 \text{ cycles}$$

PROBLEM 11 (Chapter 5)

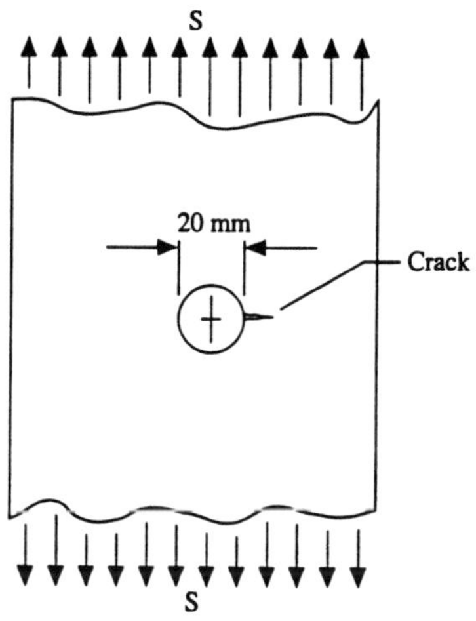

Figure P.17

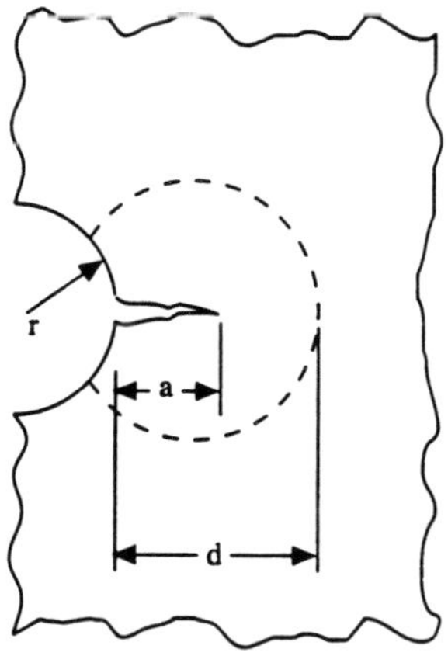

Figure P.18

Figures P.17 and P.18 (preceding page) represent a flat plate with a hole, subject to a variable load. It is made of SAE 1020 steel with the following properties:

fracture toughness	$K_{Ic} = 120\ \text{MPa}\sqrt{\text{m}}$
threshold stress intensity factor	$K_{th} = 6.0\ \text{MPa}\sqrt{\text{m}}$
constant factor in Equation (5.85)	$C = 6.9 \times 10^{-12}$
crack growth exponent in Equation (5.85)	$n = 3.0$

The variable load equals:

$S = \pm 230$ MPa

The plate has a small crack next to the hole of the length:

$a_i = 0.5\ \text{mm} = 0.0005\ \text{m}$

Required:

(a) Based on the above, predict a danger of crack propagation.
(b) What is the number of cycles to the final fracture?

SOLUTION

To arrive at the proper answer, the compression load fluctuation is ignored and it is assumed that the stress fluctuation has the form:

$S = 0\ \text{to}\ 230$ MPa

(a) At the onset of crack growth, see Figure 5.19, the threshold value of stress intensity factor must equal:

$$K_{th} = 1.12 \times \sigma \times \sqrt{\pi a} = 1.12 \times 230\sqrt{\pi a} = 6.0\,\text{MPa}\sqrt{\text{m}}$$

see Equation (5.22). Conseqently the initial length of a crack should be:

$$a = \frac{1}{\pi}\left(\frac{6.0}{1.12 \times 230}\right)^2 = 0.00017\ \text{m}$$

The given length is larger, therefore under the fluctuating loading the crack will continue to grow.

(b) It is assumed a two-step crack propagation: initially within the zone influenced strongly by the hole, see Figure 8.6 and then beyond it. The size of the zone, as per Equation (5.25) equals:

$$d = 0.13\sqrt{20 \times 10} = 1.838 \text{ mm} = 0.001838 \text{ m}$$

In the zone under influence of the hole the stress concentration factor equals:

$$K_t = 3.0$$

see Figure 3.30. The stress intensity factor, as per Equation (5.26) is:

$$K_I = 1.12\ K_t S\sqrt{\pi a}$$

The effective fluctuating stress is:

$$S' = K_t S = 3.0 \times 230 = 690 \text{ MPa}$$

The number of cycles, N_1, for the crack to pass over the zone equals:

$$N_1 = \frac{1}{C\left(\alpha K_t S\sqrt{\pi}\right)^n}\int_{a_i}^{d}\frac{da}{a^{n/2}} = \frac{2}{(3-2)\times 6.9\times 10^{-12}\times\left(1.12\times 690\sqrt{\pi}\right)^3}\left(\frac{1}{\sqrt{0.0005}} - \frac{1}{0.001838}\right) = 2413$$

The number of cycles, N_2, from the end of the zone to the point of the final fracture is expressed by the equation:

$$N_2 = \frac{1}{C\left(\alpha S\sqrt{\pi}\right)^n}\int_{D+d}^{D+a_f}\frac{da}{a^{n/2}}$$

The final crack length equals:

$$a_f = \frac{1}{\pi}\left(\frac{120}{1.12\times 230}\right)^2 = 0.0691 \text{ m}$$

resulting in the number of cycles:

$$N_2 = \frac{2}{(3-2)\times 6.9\times 10^{-12}\left(1.12\times 230\times\sqrt{\pi}\right)^3}\left(\frac{1}{\sqrt{0.021838}} - \frac{1}{\sqrt{0.0891}}\right) = 10{,}405$$

The total number of loading cycles from the beginning till the occurrence of the final fracture equals:

$$N = N_1 + N_2 = 12{,}818$$

PROBLEM 12 (Chapter 5)

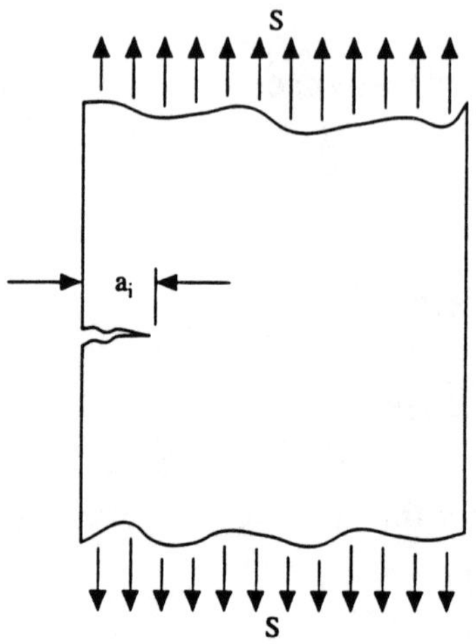

Figure P.19

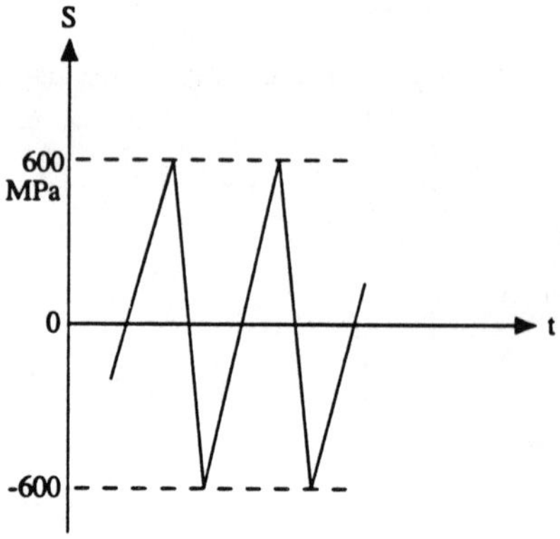

Figure P.20

Figures P.19 and P.20 (preceding page) represent a flat plate made of SAE 1020 steel, subject to a variable load. The plate has the following properties:

fracture toughness	$K_{Ic} = 120 \text{ MPa}\sqrt{\text{m}}$
threshold stress intensity factor	$K_{th} = 6.0 \text{ MPa}\sqrt{\text{m}}$
constant factor in Equation (5.85)	$C = 6.9 \times 10^{-12}$
crack growth exponent in Equation (5.85)	$n = 3.0$

The plate had a small crack at the edge, as shown in the figure. The variable load equals:

$S = \pm 220 \text{ MPa}$

The plate failed after 20,000 cycles.

Required:

(a) Compute the length of the original crack.
(b) Compute the permissible load to avoid the failure.

SOLUTION

To arrive at the proper answer, the compression during load fluctuation is ignored and it is assumed that the stress fluctuation has the form:

$S = 0 \text{ to } 220 \text{ MPa}$

(a) The number of cycles from the beginning of loading to the final fracture is:

$$20{,}000 = \frac{2}{(3-2)\times 6.9\times 10^{-12}\times\left(1.12\times 220\times\sqrt{\pi}\right)^{3}}\left(\frac{1}{\sqrt{a_i}}-\frac{1}{\sqrt{a_f}}\right)$$

see Equation (5.92). The crack length at the fracture is:

$$a_f = \frac{1}{\pi}\left(\frac{K_{Ic}}{\alpha S}\right)^2 = \frac{1}{\pi}\left(\frac{120}{1.12\times 220}\right)^2 = 0.0755 \text{ m}$$

Hence,

$$20{,}000 = \frac{2}{(3-2)\times 6.9\times 10^{-12}\times\left(1.12\times 220\times\sqrt{\pi}\right)^{3}}\left(\frac{1}{\sqrt{a_i}}-\frac{1}{\sqrt{0.0755}}\right)$$

The length of the original crack is:

$a_i = 0.0113$ m

(b) To guard against crack propagation, the stress intensity factor should be below the threshold value of 6.0 $\text{MPa}\sqrt{\text{m}}$. Assume the stress intensity factor to be:

$K_I = 5.0\ \text{MPa}\sqrt{\text{m}}$

Then,

$5.0 = 1.12 \times S \times \sqrt{\pi \times 0.0113}$

This gives the highest permissible stress:

$S = 23.7$ MPa

PROBLEM 13 (Chapter 7)

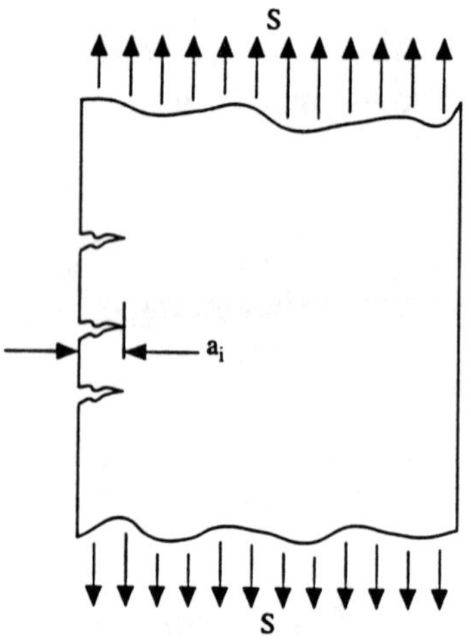

Figure P.21

Figure P.21 above represents a flat plate made of A-441 steel subject to a variable load:

$S = 0\,\text{to}\,300$ MPa

The edge of the plate has small cracks.

$a_i \leq 0.2$ mm

Discuss the danger of crack propagation. Could the danger be avoided by shot-peening?

SOLUTION

The threshold value of stress intensity factor for steel is:

$K_{th} = 6.0 \text{ MPa}\sqrt{\text{m}}$

The critical length of a crack that starts to propagate is:

$$a = \frac{1}{\pi}\left(\frac{K_{th}}{1.12S}\right)^2 = \frac{1}{\pi}\left(\frac{6.0}{1.12 \times 300}\right)^2 = 0.0001 \text{ m}$$

The existing cracks in the plate are larger than the critical length, which proves that a danger of crack propagation exists.

Assume a shot peening process with residual stresses

$\sigma_r = -200$ MPa

(Section 7.1.2). The stress intensity factor at the cracks equals:

$$K_I = 1.12\sigma\sqrt{\pi a} = 1.12(300 - 200)\sqrt{\pi \times 0.0002} = 2.8 \text{ MPa}\sqrt{\text{m}}$$

below the threshold value. Thus the crack propagation will be avoided by shot-peening.

PROBLEM 14

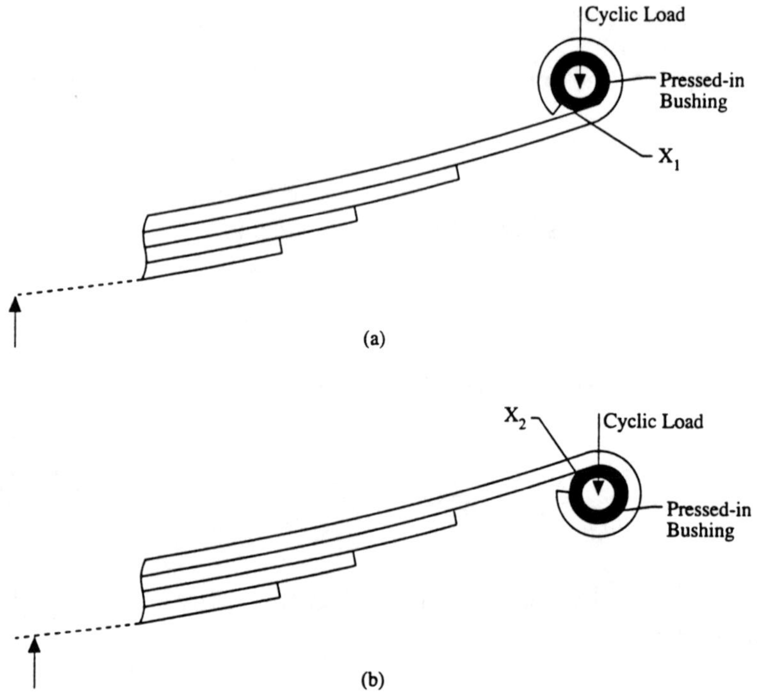

Figure P.22

The above Figure P.22 shows two leaf springs where the spring eyes have curvatures in different directions with a pressed-in bushing. Discuss in which of the two springs a fatigue failure will occur first.

SOLUTION

Both leaf springs are subject to a cyclic load that induces cyclic bending stresses in each leaf. In both cases the top leaf with the spring eye is subject to both the static tension stresses caused by the pressed-in bushing and cyclic stresses caused by working load. However, at the critical points (X_1 and X_2) the cyclic stresses have different directions.

In the spring with a turned-up eye, at point X_1 the static tension stressses are added to the cyclic tension stresses from the working load, resulting in a large combined tension stress. In the spring with a turned-down eye, at point X_2 the static tension stresses are added to the cyclic compression stresses, resulting in a lower combined stress and longer fatigue life.

Figure P.23 shows the stress conditions for both cases in a constant life diagram.

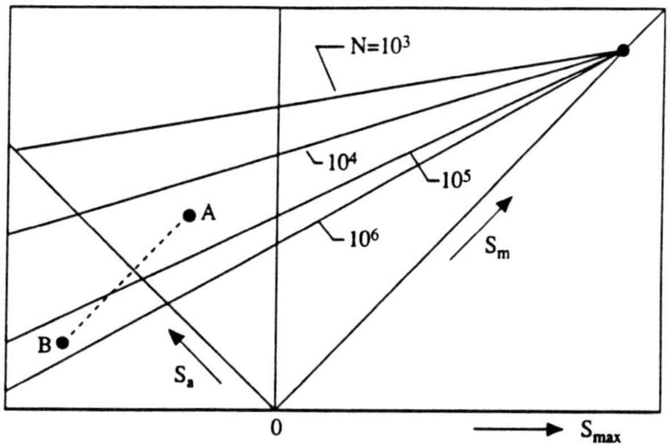

Figure P.23